THIS BOOK BELONG TO

Copyright © Teresa Rother
All rights reserved. No part of this publication may be reproduced, distributed, or transmitted in any form or by any means, including photocopy, recording, or other electronic or mechanical methods.

DEDICATION

This Beekeeping Log Book is dedicated to bee farmers and hobbyists who want to manage and track their hives, record data for the colonies and queens and maintain the health of their hive.

You are my inspiration for producing this book and I'm honored to be a part of helping you manage and retain important information regarding your apiary.

HOW TO USE THIS BOOK

This Beekeeping Log Book will help you record, collect, and organize your information in an easy to use format.

Here are examples of information for you to fill in and write the details of your operation.

Fill in the following information:

1. Hive Details- date, time, hive ID number, colony name, and yard
2. Weather Conditions- checklist for conditions
3. How Much- honey, brood, space
4. Colony Temperament- calm, nervous, aggressive
5. Colony Population- low, normal, crowded
6. Brood Pattern, Eggs, Larvae
7. Food Stores (Honey and Pollen)- high, average, low, near broad
8. Queen Observation Checklist- yes/no, marked, color
9. Queens Cell Present- capped, uncapped
10. Cell- emergency, swarm, supersedure
11. Pest and Disease Checklist
12. Space to write- treatments, feedings, to-do list/notes, and supplies
13. Colony Set up- space to map out your hives

DATE	TIME	HIVE ID NUMBER
COLONY NAME		YARD

WEATHER CONDITIONS

☀ ☁ ☁ 🌧 ⛈ ❄
☐ ☐ ☐ ☐ ☐ ☐

HOW MUCH

HONEY	BROOD	SPACE

COLONY TEMPERAMENT	COLONY POPULATION
☐ CALM ☐ NERVOUS ☐ AGGRESSIVE	☐ LOW ☐ NORMAL ☐ CROWDED

BROOD PATTERN		EGGS		LARVAE	
TIGHT	SPOTTY	YES	NO	YES	NO

FOOD STORES

HONEY	HIGH	AVERAGE	LOW	NEAR BROOD
POLLEN	HIGH	AVERAGE	LOW	NEAR BROOD

QUEEN OBSERVATION

YES ☐ NO ☐	MARKED- YES ☐ NO ☐	COLOR- W Y R G B

QUEENS CELLS PRESENT		CAPPED		UNCAPPED	
YES	NO	YES	NO	YES	NO

CELL

EMERGENCY	SWARM	SUPERSEDURE

PESTS AND DISEASE

VARROA	ANTS	WAX MOTH
MITES	STARVATION	MOLD
DYSENTARY	SMALL HIVE BEETE	BOLD BROOD
ODOR	BUCKSHOT BROOD	DEAD BEES
OTHER	OTHER	OTHER

TREATMENT	FEEDINGS

TO DO/NOTES	SUPPLIES

DATE	TIME	HIVE ID NUMBER
COLONY NAME		YARD

WEATHER CONDITIONS

☀️ ☁️ ☁️ 🌧️ ⛈️ ❄️
☐ ☐ ☐ ☐ ☐ ☐

HOW MUCH

HONEY	BROOD	SPACE

COLONY TEMPERAMENT	COLONY POPULATION
☐ CALM ☐ NERVOUS ☐ AGGRESSIVE	☐ LOW ☐ NORMAL ☐ CROWDED

BROOD PATTERN		EGGS		LARVAE	
TIGHT	SPOTTY	YES	NO	YES	NO

FOOD STORES

HONEY	HIGH	AVERAGE	LOW	NEAR BROOD
POLLEN	HIGH	AVERAGE	LOW	NEAR BROOD

QUEEN OBSERVATION

YES ☐ NO ☐   MARKED- YES ☐ NO ☐   COLOR- W Y R G B

QUEENS CELLS PRESENT		CAPPED		UNCAPPED	
YES	NO	YES	NO	YES	NO

CELL

EMERGENCY	SWARM	SUPERSEDURE

PESTS AND DISEASE

VARROA	ANTS	WAX MOTH
MITES	STARVATION	MOLD
DYSENTARY	SMALL HIVE BEETE	BOLD BROOD
ODOR	BUCKSHOT BROOD	DEAD BEES
OTHER	OTHER	OTHER

TREATMENT	FEEDINGS

TO DO/NOTES	SUPPLIES

DATE	TIME	HIVE ID NUMBER
COLONY NAME		YARD

WEATHER CONDITIONS

☀️ ⛅ ☁️ 🌧️ ⛈️ ❄️
◯ ◯ ◯ ◯ ◯ ◯

HOW MUCH

HONEY	BROOD	SPACE

COLONY TEMPERAMENT	COLONY POPULATION
☐ CALM ☐ NERVOUS ☐ AGGRESSIVE	☐ LOW ☐ NORMAL ☐ CROWDED

BROOD PATTERN		EGGS		LARVAE	
TIGHT	SPOTTY	YES	NO	YES	NO

FOOD STORES

HONEY	HIGH	AVERAGE	LOW	NEAR BROOD
POLLEN	HIGH	AVERAGE	LOW	NEAR BROOD

QUEEN OBSERVATION

YES ☐ NO ☐   MARKED- YES ☐ NO ☐   COLOR- W Y R G B

QUEENS CELLS PRESENT		CAPPED		UNCAPPED	
YES	NO	YES	NO	YES	NO

CELL

EMERGENCY	SWARM	SUPERSEDURE

PESTS AND DISEASE

VARROA	ANTS	WAX MOTH
MITES	STARVATION	MOLD
DYSENTARY	SMALL HIVE BEETE	BOLD BROOD
ODOR	BUCKSHOT BROOD	DEAD BEES
OTHER	OTHER	OTHER

TREATMENT	FEEDINGS

TO DO/NOTES	SUPPLIES

DATE	TIME	HIVE ID NUMBER
COLONY NAME		YARD

WEATHER CONDITIONS

☀️ ☁️ ☁️ 🌧️ ⛈️ ❄️
☐ ☐ ☐ ☐ ☐ ☐

HOW MUCH

HONEY	BROOD	SPACE

COLONY TEMPERAMENT	COLONY POPULATION
☐ CALM ☐ NERVOUS ☐ AGGRESSIVE	☐ LOW ☐ NORMAL ☐ CROWDED

BROOD PATTERN		EGGS		LARVAE	
TIGHT	SPOTTY	YES	NO	YES	NO

FOOD STORES

HONEY	HIGH	AVERAGE	LOW	NEAR BROOD
POLLEN	HIGH	AVERAGE	LOW	NEAR BROOD

QUEEN OBSERVATION

YES ☐   NO ☐   MARKED- YES ☐ NO ☐   COLOR- [W] [Y] [R] [G] [B]

QUEENS CELLS PRESENT		CAPPED		UNCAPPED	
YES	NO	YES	NO	YES	NO

CELL

EMERGENCY	SWARM	SUPERSEDURE

PESTS AND DISEASE

VARROA	ANTS	WAX MOTH
MITES	STARVATION	MOLD
DYSENTARY	SMALL HIVE BEETE	BOLD BROOD
ODOR	BUCKSHOT BROOD	DEAD BEES
OTHER	OTHER	OTHER

TREATMENT	FEEDINGS

TO DO/NOTES	SUPPLIES

DATE	TIME	HIVE ID NUMBER
COLONY NAME		YARD

WEATHER CONDITIONS

☀️ ⛅ ☁️ 🌧️ ⛈️ ❄️
☐ ☐ ☐ ☐ ☐ ☐

HOW MUCH

HONEY	BROOD	SPACE

COLONY TEMPERAMENT
☐ CALM ☐ NERVOUS ☐ AGGRESSIVE

COLONY POPULATION
☐ LOW ☐ NORMAL ☐ CROWDED

BROOD PATTERN		EGGS		LARVAE	
TIGHT	SPOTTY	YES	NO	YES	NO

FOOD STORES

HONEY	HIGH	AVERAGE	LOW	NEAR BROOD
POLLEN	HIGH	AVERAGE	LOW	NEAR BROOD

QUEEN OBSERVATION

YES ☐   NO ☐   MARKED- YES ☐   NO ☐   COLOR- [W] [Y] [R] [G] [B]

QUEENS CELLS PRESENT		CAPPED		UNCAPPED	
YES	NO	YES	NO	YES	NO

CELL

EMERGENCY	SWARM	SUPERSEDURE

PESTS AND DISEASE

VARROA	ANTS	WAX MOTH
MITES	STARVATION	MOLD
DYSENTARY	SMALL HIVE BEETE	BOLD BROOD
ODOR	BUCKSHOT BROOD	DEAD BEES
OTHER	OTHER	OTHER

TREATMENT	FEEDINGS

TO DO/NOTES	SUPPLIES

DATE	TIME	HIVE ID NUMBER
COLONY NAME		YARD

WEATHER CONDITIONS

☀️ ☁️ ☁️ 🌧️ ⛈️ ❄️
☐ ☐ ☐ ☐ ☐ ☐

HOW MUCH

HONEY	BROOD	SPACE

COLONY TEMPERAMENT	COLONY POPULATION
☐ CALM ☐ NERVOUS ☐ AGGRESSIVE	☐ LOW ☐ NORMAL ☐ CROWDED

BROOD PATTERN		EGGS		LARVAE	
TIGHT	SPOTTY	YES	NO	YES	NO

FOOD STORES

HONEY	HIGH	AVERAGE	LOW	NEAR BROOD
POLLEN	HIGH	AVERAGE	LOW	NEAR BROOD

QUEEN OBSERVATION

YES ☐ NO ☐	MARKED- YES ☐ NO ☐	COLOR- W Y R G B

QUEENS CELLS PRESENT		CAPPED		UNCAPPED	
YES	NO	YES	NO	YES	NO

CELL

EMERGENCY	SWARM	SUPERSEDURE

PESTS AND DISEASE

VARROA	ANTS	WAX MOTH
MITES	STARVATION	MOLD
DYSENTARY	SMALL HIVE BEETE	BOLD BROOD
ODOR	BUCKSHOT BROOD	DEAD BEES
OTHER	OTHER	OTHER

TREATMENT	FEEDINGS

TO DO/NOTES	SUPPLIES

DATE		TIME	HIVE ID NUMBER
COLONY NAME		YARD	

WEATHER CONDITIONS

☀️ ☁️ ☁️ 🌧️ ⛈️ ❄️
☐ ☐ ☐ ☐ ☐ ☐

HOW MUCH

HONEY	BROOD	SPACE

COLONY TEMPERAMENT	COLONY POPULATION
☐ CALM ☐ NERVOUS ☐ AGGRESSIVE	☐ LOW ☐ NORMAL ☐ CROWDED

BROOD PATTERN		EGGS		LARVAE	
TIGHT	SPOTTY	YES	NO	YES	NO

FOOD STORES

HONEY	HIGH	AVERAGE	LOW	NEAR BROOD
POLLEN	HIGH	AVERAGE	LOW	NEAR BROOD

QUEEN OBSERVATION

YES ☐   NO ☐   MARKED- YES ☐ NO ☐   COLOR- W Y R G B

QUEENS CELLS PRESENT		CAPPED		UNCAPPED	
YES	NO	YES	NO	YES	NO

CELL

EMERGENCY	SWARM	SUPERSEDURE

PESTS AND DISEASE

VARROA	ANTS	WAX MOTH
MITES	STARVATION	MOLD
DYSENTARY	SMALL HIVE BEETE	BOLD BROOD
ODOR	BUCKSHOT BROOD	DEAD BEES
OTHER	OTHER	OTHER

TREATMENT	FEEDINGS

TO DO/NOTES	SUPPLIES

DATE	TIME	HIVE ID NUMBER
COLONY NAME		YARD

WEATHER CONDITIONS

☀️ ☁️ ☁️ 🌧️ ⛈️ ❄️
☐ ☐ ☐ ☐ ☐ ☐

HOW MUCH

HONEY	BROOD	SPACE

COLONY TEMPERAMENT	COLONY POPULATION
☐ CALM ☐ NERVOUS ☐ AGGRESSIVE	☐ LOW ☐ NORMAL ☐ CROWDED

BROOD PATTERN		EGGS		LARVAE	
TIGHT	SPOTTY	YES	NO	YES	NO

FOOD STORES

HONEY	HIGH	AVERAGE	LOW	NEAR BROOD
POLLEN	HIGH	AVERAGE	LOW	NEAR BROOD

QUEEN OBSERVATION

YES ☐ NO ☐   MARKED- YES ☐ NO ☐   COLOR- W Y R G B

QUEENS CELLS PRESENT		CAPPED		UNCAPPED	
YES	NO	YES	NO	YES	NO

CELL

EMERGENCY	SWARM	SUPERSEDURE

PESTS AND DISEASE

VARROA	ANTS	WAX MOTH
MITES	STARVATION	MOLD
DYSENTARY	SMALL HIVE BEETE	BOLD BROOD
ODOR	BUCKSHOT BROOD	DEAD BEES
OTHER	OTHER	OTHER

TREATMENT	FEEDINGS

TO DO/NOTES	SUPPLIES

DATE	TIME	HIVE ID NUMBER
COLONY NAME		YARD

WEATHER CONDITIONS

☀️ ⛅ ☁️ 🌧️ ⛈️ ❄️
☐ ☐ ☐ ☐ ☐ ☐

HOW MUCH

HONEY	BROOD	SPACE

COLONY TEMPERAMENT	COLONY POPULATION
☐ CALM ☐ NERVOUS ☐ AGGRESSIVE	☐ LOW ☐ NORMAL ☐ CROWDED

BROOD PATTERN		EGGS		LARVAE	
TIGHT	SPOTTY	YES	NO	YES	NO

FOOD STORES

HONEY	HIGH	AVERAGE	LOW	NEAR BROOD
POLLEN	HIGH	AVERAGE	LOW	NEAR BROOD

QUEEN OBSERVATION

YES ☐ NO ☐   MARKED- YES ☐ NO ☐   COLOR- W Y R G B

QUEENS CELLS PRESENT		CAPPED		UNCAPPED	
YES	NO	YES	NO	YES	NO

CELL

EMERGENCY	SWARM	SUPERSEDURE

PESTS AND DISEASE

VARROA	ANTS	WAX MOTH
MITES	STARVATION	MOLD
DYSENTARY	SMALL HIVE BEETE	BOLD BROOD
ODOR	BUCKSHOT BROOD	DEAD BEES
OTHER	OTHER	OTHER

TREATMENT	FEEDINGS

TO DO/NOTES	SUPPLIES

DATE	TIME	HIVE ID NUMBER
COLONY NAME		YARD

WEATHER CONDITIONS

☀️ ☁️ ⛅ 🌧️ ⛈️ ❄️
☐ ☐ ☐ ☐ ☐ ☐

HOW MUCH

HONEY	BROOD	SPACE

COLONY TEMPERAMENT	COLONY POPULATION
☐ CALM ☐ NERVOUS ☐ AGGRESSIVE	☐ LOW ☐ NORMAL ☐ CROWDED

BROOD PATTERN		EGGS		LARVAE	
TIGHT	SPOTTY	YES	NO	YES	NO

FOOD STORES

HONEY	HIGH	AVERAGE	LOW	NEAR BROOD
POLLEN	HIGH	AVERAGE	LOW	NEAR BROOD

QUEEN OBSERVATION

YES ☐ NO ☐ MARKED- YES ☐ NO ☐ COLOR- W Y R G B

QUEENS CELLS PRESENT		CAPPED		UNCAPPED	
YES	NO	YES	NO	YES	NO

CELL

EMERGENCY	SWARM	SUPERSEDURE

PESTS AND DISEASE

VARROA	ANTS	WAX MOTH
MITES	STARVATION	MOLD
DYSENTARY	SMALL HIVE BEETE	BOLD BROOD
ODOR	BUCKSHOT BROOD	DEAD BEES
OTHER	OTHER	OTHER

TREATMENT	FEEDINGS

TO DO/NOTES	SUPPLIES

DATE	TIME	HIVE ID NUMBER
COLONY NAME		YARD

WEATHER CONDITIONS

☀️ ⛅ ☁️ 🌧️ ⛈️ ❄️
○ ○ ○ ○ ○ ○

HOW MUCH

HONEY	BROOD	SPACE

COLONY TEMPERAMENT	COLONY POPULATION
☐ CALM ☐ NERVOUS ☐ AGGRESSIVE	☐ LOW ☐ NORMAL ☐ CROWDED

BROOD PATTERN		EGGS		LARVAE	
TIGHT	SPOTTY	YES	NO	YES	NO

FOOD STORES

HONEY	HIGH	AVERAGE	LOW	NEAR BROOD
POLLEN	HIGH	AVERAGE	LOW	NEAR BROOD

QUEEN OBSERVATION

YES ☐ NO ☐   MARKED- YES ☐ NO ☐   COLOR- W Y R G B

QUEENS CELLS PRESENT		CAPPED		UNCAPPED	
YES	NO	YES	NO	YES	NO

CELL

EMERGENCY	SWARM	SUPERSEDURE

PESTS AND DISEASE

VARROA	ANTS	WAX MOTH
MITES	STARVATION	MOLD
DYSENTARY	SMALL HIVE BEETE	BOLD BROOD
ODOR	BUCKSHOT BROOD	DEAD BEES
OTHER	OTHER	OTHER

TREATMENT	FEEDINGS

TO DO/NOTES	SUPPLIES

DATE	TIME	HIVE ID NUMBER
COLONY NAME		YARD

WEATHER CONDITIONS

☀️ ☁️ ☁️ 🌧️ ⛈️ ❄️
☐ ☐ ☐ ☐ ☐ ☐

HOW MUCH

HONEY	BROOD	SPACE

COLONY TEMPERAMENT	COLONY POPULATION
☐ CALM ☐ NERVOUS ☐ AGGRESSIVE	☐ LOW ☐ NORMAL ☐ CROWDED

BROOD PATTERN		EGGS		LARVAE	
TIGHT	SPOTTY	YES	NO	YES	NO

FOOD STORES

HONEY	HIGH	AVERAGE	LOW	NEAR BROOD
POLLEN	HIGH	AVERAGE	LOW	NEAR BROOD

QUEEN OBSERVATION

YES ☐ NO ☐ MARKED- YES ☐ NO ☐ COLOR- W Y R G B

QUEENS CELLS PRESENT		CAPPED		UNCAPPED	
YES	NO	YES	NO	YES	NO

CELL

EMERGENCY	SWARM	SUPERSEDURE

PESTS AND DISEASE

VARROA	ANTS	WAX MOTH
MITES	STARVATION	MOLD
DYSENTARY	SMALL HIVE BEETE	BOLD BROOD
ODOR	BUCKSHOT BROOD	DEAD BEES
OTHER	OTHER	OTHER

TREATMENT	FEEDINGS

TO DO/NOTES	SUPPLIES

DATE	TIME	HIVE ID NUMBER
COLONY NAME		YARD

WEATHER CONDITIONS

☀️ ☁️ ⛅ 🌧️ ⛈️ ❄️
☐ ☐ ☐ ☐ ☐ ☐

HOW MUCH

HONEY	BROOD	SPACE

COLONY TEMPERAMENT	COLONY POPULATION
☐ CALM ☐ NERVOUS ☐ AGGRESSIVE	☐ LOW ☐ NORMAL ☐ CROWDED

BROOD PATTERN		EGGS		LARVAE	
TIGHT	SPOTTY	YES	NO	YES	NO

FOOD STORES

HONEY	HIGH	AVERAGE	LOW	NEAR BROOD
POLLEN	HIGH	AVERAGE	LOW	NEAR BROOD

QUEEN OBSERVATION

YES ☐ NO ☐	MARKED- YES ☐ NO ☐	COLOR- W Y R G B

QUEENS CELLS PRESENT		CAPPED		UNCAPPED	
YES	NO	YES	NO	YES	NO

CELL

EMERGENCY	SWARM	SUPERSEDURE

PESTS AND DISEASE

VARROA	ANTS	WAX MOTH
MITES	STARVATION	MOLD
DYSENTARY	SMALL HIVE BEETE	BOLD BROOD
ODOR	BUCKSHOT BROOD	DEAD BEES
OTHER	OTHER	OTHER

TREATMENT	FEEDINGS

TO DO/NOTES	SUPPLIES

DATE	TIME	HIVE ID NUMBER
COLONY NAME		YARD

WEATHER CONDITIONS

☀️ ○ ⛅ ○ ☁️ ○ 🌧️ ○ ⛈️ ○ ❄️ ○

HOW MUCH

HONEY	BROOD	SPACE

COLONY TEMPERAMENT	COLONY POPULATION
☐ CALM ☐ NERVOUS ☐ AGGRESSIVE	☐ LOW ☐ NORMAL ☐ CROWDED

BROOD PATTERN		EGGS		LARVAE	
TIGHT	SPOTTY	YES	NO	YES	NO

FOOD STORES

HONEY	HIGH	AVERAGE	LOW	NEAR BROOD
POLLEN	HIGH	AVERAGE	LOW	NEAR BROOD

QUEEN OBSERVATION

YES ☐   NO ☐   MARKED- YES ☐ NO ☐   COLOR- ☐W ☐Y ☐R ☐G ☐B

QUEENS CELLS PRESENT		CAPPED		UNCAPPED	
YES	NO	YES	NO	YES	NO

CELL

EMERGENCY	SWARM	SUPERSEDURE

PESTS AND DISEASE

VARROA	ANTS	WAX MOTH
MITES	STARVATION	MOLD
DYSENTARY	SMALL HIVE BEETE	BOLD BROOD
ODOR	BUCKSHOT BROOD	DEAD BEES
OTHER	OTHER	OTHER

TREATMENT	FEEDINGS

TO DO/NOTES	SUPPLIES

DATE	TIME	HIVE ID NUMBER
COLONY NAME		YARD

WEATHER CONDITIONS

☀️ ☁️ ☁️ 🌧️ ⛈️ ❄️
☐ ☐ ☐ ☐ ☐ ☐

HOW MUCH

HONEY	BROOD	SPACE

COLONY TEMPERAMENT	COLONY POPULATION
☐ CALM ☐ NERVOUS ☐ AGGRESSIVE	☐ LOW ☐ NORMAL ☐ CROWDED

BROOD PATTERN		EGGS		LARVAE	
TIGHT	SPOTTY	YES	NO	YES	NO

FOOD STORES

HONEY	HIGH	AVERAGE	LOW	NEAR BROOD
POLLEN	HIGH	AVERAGE	LOW	NEAR BROOD

QUEEN OBSERVATION

YES ☐ NO ☐   MARKED- YES ☐ NO ☐   COLOR- W Y R G B

QUEENS CELLS PRESENT		CAPPED		UNCAPPED	
YES	NO	YES	NO	YES	NO

CELL

EMERGENCY	SWARM	SUPERSEDURE

PESTS AND DISEASE

VARROA	ANTS	WAX MOTH
MITES	STARVATION	MOLD
DYSENTARY	SMALL HIVE BEETE	BOLD BROOD
ODOR	BUCKSHOT BROOD	DEAD BEES
OTHER	OTHER	OTHER

TREATMENT	FEEDINGS

TO DO/NOTES	SUPPLIES

DATE	TIME	HIVE ID NUMBER
COLONY NAME		YARD

WEATHER CONDITIONS

☀️ ☁️ ☁️ 🌧️ ⛈️ ❄️
☐ ☐ ☐ ☐ ☐ ☐

HOW MUCH

HONEY	BROOD	SPACE

COLONY TEMPERAMENT	COLONY POPULATION
☐ CALM ☐ NERVOUS ☐ AGGRESSIVE	☐ LOW ☐ NORMAL ☐ CROWDED

BROOD PATTERN		EGGS		LARVAE	
TIGHT	SPOTTY	YES	NO	YES	NO

FOOD STORES

HONEY	HIGH	AVERAGE	LOW	NEAR BROOD
POLLEN	HIGH	AVERAGE	LOW	NEAR BROOD

QUEEN OBSERVATION

YES ☐ NO ☐	MARKED- YES ☐ NO ☐	COLOR- W Y R G B

QUEENS CELLS PRESENT		CAPPED		UNCAPPED	
YES	NO	YES	NO	YES	NO

CELL

EMERGENCY	SWARM	SUPERSEDURE

PESTS AND DISEASE

VARROA	ANTS	WAX MOTH
MITES	STARVATION	MOLD
DYSENTARY	SMALL HIVE BEETE	BOLD BROOD
ODOR	BUCKSHOT BROOD	DEAD BEES
OTHER	OTHER	OTHER

TREATMENT	FEEDINGS

TO DO/NOTES	SUPPLIES

DATE		TIME	HIVE ID NUMBER
COLONY NAME		YARD	

WEATHER CONDITIONS

☀️ ☁️ ☁️ 🌧️ ⛈️ ❄️
○ ○ ○ ○ ○ ○

HOW MUCH

HONEY	BROOD	SPACE

COLONY TEMPERAMENT	COLONY POPULATION
☐ CALM ☐ NERVOUS ☐ AGGRESSIVE	☐ LOW ☐ NORMAL ☐ CROWDED

BROOD PATTERN		EGGS		LARVAE	
TIGHT	SPOTTY	YES	NO	YES	NO

FOOD STORES

HONEY	HIGH	AVERAGE	LOW	NEAR BROOD
POLLEN	HIGH	AVERAGE	LOW	NEAR BROOD

QUEEN OBSERVATION

YES ☐ NO ☐   MARKED- YES ☐ NO ☐   COLOR- W Y R G B

QUEENS CELLS PRESENT		CAPPED		UNCAPPED	
YES	NO	YES	NO	YES	NO

CELL

EMERGENCY	SWARM	SUPERSEDURE

PESTS AND DISEASE

VARROA	ANTS	WAX MOTH
MITES	STARVATION	MOLD
DYSENTARY	SMALL HIVE BEETE	BOLD BROOD
ODOR	BUCKSHOT BROOD	DEAD BEES
OTHER	OTHER	OTHER

TREATMENT	FEEDINGS

TO DO/NOTES	SUPPLIES

DATE	TIME	HIVE ID NUMBER
COLONY NAME		YARD

WEATHER CONDITIONS

☀️	⛅	☁️	🌧️	⛈️	❄️
☐	☐	☐	☐	☐	☐

HOW MUCH

HONEY	BROOD	SPACE

COLONY TEMPERAMENT	COLONY POPULATION
☐ CALM ☐ NERVOUS ☐ AGGRESSIVE	☐ LOW ☐ NORMAL ☐ CROWDED

BROOD PATTERN		EGGS		LARVAE	
TIGHT	SPOTTY	YES	NO	YES	NO

FOOD STORES

HONEY	HIGH	AVERAGE	LOW	NEAR BROOD
POLLEN	HIGH	AVERAGE	LOW	NEAR BROOD

QUEEN OBSERVATION

YES ☐ NO ☐	MARKED- YES ☐ NO ☐	COLOR- ☐W ☐Y ☐R ☐G ☐B

QUEENS CELLS PRESENT		CAPPED		UNCAPPED	
YES	NO	YES	NO	YES	NO

CELL

EMERGENCY	SWARM	SUPERSEDURE

PESTS AND DISEASE

VARROA	ANTS	WAX MOTH
MITES	STARVATION	MOLD
DYSENTARY	SMALL HIVE BEETE	BOLD BROOD
ODOR	BUCKSHOT BROOD	DEAD BEES
OTHER	OTHER	OTHER

TREATMENT	FEEDINGS

TO DO/NOTES	SUPPLIES

DATE	TIME	HIVE ID NUMBER
COLONY NAME		YARD

WEATHER CONDITIONS

☀ ○ ⛅ ○ ☁ ○ 🌧 ○ ⛈ ○ ❄ ○

HOW MUCH

HONEY	BROOD	SPACE

COLONY TEMPERAMENT	COLONY POPULATION
☐ CALM ☐ NERVOUS ☐ AGGRESSIVE	☐ LOW ☐ NORMAL ☐ CROWDED

BROOD PATTERN		EGGS		LARVAE	
TIGHT	SPOTTY	YES	NO	YES	NO

FOOD STORES

HONEY	HIGH	AVERAGE	LOW	NEAR BROOD
POLLEN	HIGH	AVERAGE	LOW	NEAR BROOD

QUEEN OBSERVATION

YES ☐   NO ☐   MARKED- YES ☐ NO ☐   COLOR- W Y R G B

QUEENS CELLS PRESENT		CAPPED		UNCAPPED	
YES	NO	YES	NO	YES	NO

CELL

EMERGENCY	SWARM	SUPERSEDURE

PESTS AND DISEASE

VARROA	ANTS	WAX MOTH
MITES	STARVATION	MOLD
DYSENTARY	SMALL HIVE BEETE	BOLD BROOD
ODOR	BUCKSHOT BROOD	DEAD BEES
OTHER	OTHER	OTHER

TREATMENT	FEEDINGS

TO DO/NOTES	SUPPLIES

DATE	TIME	HIVE ID NUMBER
COLONY NAME		YARD

WEATHER CONDITIONS

☀️ ☁️ ☁️ 🌧️ ⛈️ ❄️
☐ ☐ ☐ ☐ ☐ ☐

HOW MUCH

HONEY	BROOD	SPACE

COLONY TEMPERAMENT	COLONY POPULATION
☐ CALM ☐ NERVOUS ☐ AGGRESSIVE	☐ LOW ☐ NORMAL ☐ CROWDED

BROOD PATTERN		EGGS		LARVAE	
TIGHT	SPOTTY	YES	NO	YES	NO

FOOD STORES

HONEY	HIGH	AVERAGE	LOW	NEAR BROOD
POLLEN	HIGH	AVERAGE	LOW	NEAR BROOD

QUEEN OBSERVATION

YES ☐ NO ☐	MARKED- YES ☐ NO ☐	COLOR- W Y R G B

QUEENS CELLS PRESENT		CAPPED		UNCAPPED	
YES	NO	YES	NO	YES	NO

CELL

EMERGENCY	SWARM	SUPERSEDURE

PESTS AND DISEASE

VARROA	ANTS	WAX MOTH
MITES	STARVATION	MOLD
DYSENTARY	SMALL HIVE BEETE	BOLD BROOD
ODOR	BUCKSHOT BROOD	DEAD BEES
OTHER	OTHER	OTHER

TREATMENT	FEEDINGS

TO DO/NOTES	SUPPLIES

DATE	TIME	HIVE ID NUMBER
COLONY NAME		YARD

WEATHER CONDITIONS

☀️ ⛅ ☁️ 🌧️ ⛈️ ❄️
○ ○ ○ ○ ○ ○

HOW MUCH

HONEY	BROOD	SPACE

COLONY TEMPERAMENT	COLONY POPULATION
☐ CALM ☐ NERVOUS ☐ AGGRESSIVE	☐ LOW ☐ NORMAL ☐ CROWDED

BROOD PATTERN		EGGS		LARVAE	
TIGHT	SPOTTY	YES	NO	YES	NO

FOOD STORES

HONEY	HIGH	AVERAGE	LOW	NEAR BROOD
POLLEN	HIGH	AVERAGE	LOW	NEAR BROOD

QUEEN OBSERVATION

YES ☐ NO ☐   MARKED- YES ☐ NO ☐   COLOR- W Y R G B

QUEENS CELLS PRESENT		CAPPED		UNCAPPED	
YES	NO	YES	NO	YES	NO

CELL

EMERGENCY	SWARM	SUPERSEDURE

PESTS AND DISEASE

VARROA	ANTS	WAX MOTH
MITES	STARVATION	MOLD
DYSENTARY	SMALL HIVE BEETE	BOLD BROOD
ODOR	BUCKSHOT BROOD	DEAD BEES
OTHER	OTHER	OTHER

TREATMENT	FEEDINGS

TO DO/NOTES	SUPPLIES

DATE	TIME	HIVE ID NUMBER
COLONY NAME		YARD

WEATHER CONDITIONS

☀️ ☁️ ☁️ 🌧️ ⛈️ ❄️
☐ ☐ ☐ ☐ ☐ ☐

HOW MUCH

HONEY	BROOD	SPACE

COLONY TEMPERAMENT	COLONY POPULATION
☐ CALM ☐ NERVOUS ☐ AGGRESSIVE	☐ LOW ☐ NORMAL ☐ CROWDED

BROOD PATTERN		EGGS		LARVAE	
TIGHT	SPOTTY	YES	NO	YES	NO

FOOD STORES

HONEY	HIGH	AVERAGE	LOW	NEAR BROOD
POLLEN	HIGH	AVERAGE	LOW	NEAR BROOD

QUEEN OBSERVATION

YES ☐ NO ☐   MARKED- YES ☐ NO ☐   COLOR- W Y R G B

QUEENS CELLS PRESENT		CAPPED		UNCAPPED	
YES	NO	YES	NO	YES	NO

CELL

EMERGENCY	SWARM	SUPERSEDURE

PESTS AND DISEASE

VARROA	ANTS	WAX MOTH
MITES	STARVATION	MOLD
DYSENTARY	SMALL HIVE BEETE	BOLD BROOD
ODOR	BUCKSHOT BROOD	DEAD BEES
OTHER	OTHER	OTHER

TREATMENT	FEEDINGS

TO DO/NOTES	SUPPLIES

DATE	TIME	HIVE ID NUMBER
COLONY NAME		YARD

WEATHER CONDITIONS

☀️ ⛅ ☁️ 🌧️ ⛈️ ❄️
☐ ☐ ☐ ☐ ☐ ☐

HOW MUCH

HONEY	BROOD	SPACE

COLONY TEMPERAMENT	COLONY POPULATION
☐ CALM ☐ NERVOUS ☐ AGGRESSIVE	☐ LOW ☐ NORMAL ☐ CROWDED

BROOD PATTERN		EGGS		LARVAE	
TIGHT	SPOTTY	YES	NO	YES	NO

FOOD STORES

HONEY	HIGH	AVERAGE	LOW	NEAR BROOD
POLLEN	HIGH	AVERAGE	LOW	NEAR BROOD

QUEEN OBSERVATION

YES ☐ NO ☐ MARKED- YES ☐ NO ☐ COLOR- W Y R G B

QUEENS CELLS PRESENT		CAPPED		UNCAPPED	
YES	NO	YES	NO	YES	NO

CELL

EMERGENCY	SWARM	SUPERSEDURE

PESTS AND DISEASE

VARROA	ANTS	WAX MOTH
MITES	STARVATION	MOLD
DYSENTARY	SMALL HIVE BEETE	BOLD BROOD
ODOR	BUCKSHOT BROOD	DEAD BEES
OTHER	OTHER	OTHER

TREATMENT	FEEDINGS

TO DO/NOTES	SUPPLIES

DATE	TIME	HIVE ID NUMBER
COLONY NAME		YARD

WEATHER CONDITIONS

☀️ ☁️ ☁️ 🌧️ ⛈️ ❄️
☐ ☐ ☐ ☐ ☐ ☐

HOW MUCH

HONEY	BROOD	SPACE

COLONY TEMPERAMENT	COLONY POPULATION
☐ CALM ☐ NERVOUS ☐ AGGRESSIVE	☐ LOW ☐ NORMAL ☐ CROWDED

BROOD PATTERN		EGGS		LARVAE	
TIGHT	SPOTTY	YES	NO	YES	NO

FOOD STORES

HONEY	HIGH	AVERAGE	LOW	NEAR BROOD
POLLEN	HIGH	AVERAGE	LOW	NEAR BROOD

QUEEN OBSERVATION

YES ☐   NO ☐   MARKED- YES ☐ NO ☐   COLOR- W Y R G B

QUEENS CELLS PRESENT		CAPPED		UNCAPPED	
YES	NO	YES	NO	YES	NO

CELL

EMERGENCY	SWARM	SUPERSEDURE

PESTS AND DISEASE

VARROA	ANTS	WAX MOTH
MITES	STARVATION	MOLD
DYSENTARY	SMALL HIVE BEETE	BOLD BROOD
ODOR	BUCKSHOT BROOD	DEAD BEES
OTHER	OTHER	OTHER

TREATMENT	FEEDINGS

TO DO/NOTES	SUPPLIES

DATE	TIME	HIVE ID NUMBER
COLONY NAME		YARD

WEATHER CONDITIONS

☀️ ☁️ ☁️ 🌧️ ⛈️ ❄️
◯ ◯ ◯ ◯ ◯ ◯

HOW MUCH

HONEY	BROOD	SPACE

COLONY TEMPERAMENT	COLONY POPULATION
☐ CALM ☐ NERVOUS ☐ AGGRESSIVE	☐ LOW ☐ NORMAL ☐ CROWDED

BROOD PATTERN		EGGS		LARVAE	
TIGHT	SPOTTY	YES	NO	YES	NO

FOOD STORES

HONEY	HIGH	AVERAGE	LOW	NEAR BROOD
POLLEN	HIGH	AVERAGE	LOW	NEAR BROOD

QUEEN OBSERVATION

YES ☐ NO ☐ MARKED- YES ☐ NO ☐ COLOR- W Y R G B

QUEENS CELLS PRESENT		CAPPED		UNCAPPED	
YES	NO	YES	NO	YES	NO

CELL

EMERGENCY	SWARM	SUPERSEDURE

PESTS AND DISEASE

VARROA	ANTS	WAX MOTH
MITES	STARVATION	MOLD
DYSENTARY	SMALL HIVE BEETE	BOLD BROOD
ODOR	BUCKSHOT BROOD	DEAD BEES
OTHER	OTHER	OTHER

TREATMENT	FEEDINGS

TO DO/NOTES	SUPPLIES

DATE	TIME	HIVE ID NUMBER
COLONY NAME		YARD

WEATHER CONDITIONS

☀️ ☐   ⛅ ☐   ☁️ ☐   🌧️ ☐   ⛈️ ☐   ❄️ ☐

HOW MUCH

HONEY	BROOD	SPACE

COLONY TEMPERAMENT	COLONY POPULATION
☐ CALM ☐ NERVOUS ☐ AGGRESSIVE	☐ LOW ☐ NORMAL ☐ CROWDED

BROOD PATTERN		EGGS		LARVAE	
TIGHT	SPOTTY	YES	NO	YES	NO

FOOD STORES

HONEY	HIGH	AVERAGE	LOW	NEAR BROOD
POLLEN	HIGH	AVERAGE	LOW	NEAR BROOD

QUEEN OBSERVATION

YES ☐ NO ☐	MARKED- YES ☐ NO ☐	COLOR- W Y R G B

QUEENS CELLS PRESENT		CAPPED		UNCAPPED	
YES	NO	YES	NO	YES	NO

CELL

EMERGENCY	SWARM	SUPERSEDURE

PESTS AND DISEASE

VARROA	ANTS	WAX MOTH
MITES	STARVATION	MOLD
DYSENTARY	SMALL HIVE BEETE	BOLD BROOD
ODOR	BUCKSHOT BROOD	DEAD BEES
OTHER	OTHER	OTHER

TREATMENT	FEEDINGS

TO DO/NOTES	SUPPLIES

DATE	TIME	HIVE ID NUMBER
COLONY NAME		YARD

WEATHER CONDITIONS

☀️ ⛅ ☁️ 🌧️ ⛈️ ❄️
◯ ◯ ◯ ◯ ◯ ◯

HOW MUCH

HONEY	BROOD	SPACE

COLONY TEMPERAMENT
☐ CALM ☐ NERVOUS ☐ AGGRESSIVE

COLONY POPULATION
☐ LOW ☐ NORMAL ☐ CROWDED

BROOD PATTERN		EGGS		LARVAE	
TIGHT	SPOTTY	YES	NO	YES	NO

FOOD STORES

HONEY	HIGH	AVERAGE	LOW	NEAR BROOD
POLLEN	HIGH	AVERAGE	LOW	NEAR BROOD

QUEEN OBSERVATION

YES ☐ NO ☐   MARKED- YES ☐ NO ☐   COLOR- ☒W ☐Y ☐R ☐G ☐B

QUEENS CELLS PRESENT		CAPPED		UNCAPPED	
YES	NO	YES	NO	YES	NO

CELL

EMERGENCY	SWARM	SUPERSEDURE

PESTS AND DISEASE

VARROA	ANTS	WAX MOTH
MITES	STARVATION	MOLD
DYSENTARY	SMALL HIVE BEETE	BOLD BROOD
ODOR	BUCKSHOT BROOD	DEAD BEES
OTHER	OTHER	OTHER

TREATMENT	FEEDINGS

TO DO/NOTES	SUPPLIES

DATE	TIME	HIVE ID NUMBER
COLONY NAME		YARD

WEATHER CONDITIONS

☀️ ☐   ⛅ ☐   ☁️ ☐   🌧️ ☐   ⛈️ ☐   ❄️ ☐

HOW MUCH

HONEY	BROOD	SPACE

COLONY TEMPERAMENT	COLONY POPULATION
☐ CALM ☐ NERVOUS ☐ AGGRESSIVE	☐ LOW ☐ NORMAL ☐ CROWDED

BROOD PATTERN		EGGS		LARVAE	
TIGHT	SPOTTY	YES	NO	YES	NO

FOOD STORES

HONEY	HIGH	AVERAGE	LOW	NEAR BROOD
POLLEN	HIGH	AVERAGE	LOW	NEAR BROOD

QUEEN OBSERVATION

YES ☐   NO ☐   MARKED- YES ☐   NO ☐   COLOR- W ☐ Y ☐ R ☐ G ☐ B ☐

QUEENS CELLS PRESENT		CAPPED		UNCAPPED	
YES	NO	YES	NO	YES	NO

CELL

EMERGENCY	SWARM	SUPERSEDURE

PESTS AND DISEASE

VARROA	ANTS	WAX MOTH
MITES	STARVATION	MOLD
DYSENTARY	SMALL HIVE BEETE	BOLD BROOD
ODOR	BUCKSHOT BROOD	DEAD BEES
OTHER	OTHER	OTHER

TREATMENT	FEEDINGS

TO DO/NOTES	SUPPLIES

DATE	TIME	HIVE ID NUMBER
COLONY NAME		YARD

WEATHER CONDITIONS

☀️ ⛅ ☁️ 🌧️ ⛈️ ❄️
◯ ◯ ◯ ◯ ◯ ◯

HOW MUCH

HONEY	BROOD	SPACE

COLONY TEMPERAMENT	COLONY POPULATION
☐ CALM ☐ NERVOUS ☐ AGGRESSIVE	☐ LOW ☐ NORMAL ☐ CROWDED

BROOD PATTERN		EGGS		LARVAE	
TIGHT	SPOTTY	YES	NO	YES	NO

FOOD STORES

HONEY	HIGH	AVERAGE	LOW	NEAR BROOD
POLLEN	HIGH	AVERAGE	LOW	NEAR BROOD

QUEEN OBSERVATION

YES ☐ NO ☐	MARKED- YES ☐ NO ☐	COLOR- W Y R G B

QUEENS CELLS PRESENT		CAPPED		UNCAPPED	
YES	NO	YES	NO	YES	NO

CELL

EMERGENCY	SWARM	SUPERSEDURE

PESTS AND DISEASE

VARROA	ANTS	WAX MOTH
MITES	STARVATION	MOLD
DYSENTARY	SMALL HIVE BEETE	BOLD BROOD
ODOR	BUCKSHOT BROOD	DEAD BEES
OTHER	OTHER	OTHER

TREATMENT	FEEDINGS

TO DO/NOTES	SUPPLIES

DATE	TIME	HIVE ID NUMBER
COLONY NAME		YARD

WEATHER CONDITIONS

☀️ ⛅ ☁️ 🌧️ ⛈️ ❄️
☐ ☐ ☐ ☐ ☐ ☐

HOW MUCH

HONEY	BROOD	SPACE

COLONY TEMPERAMENT	COLONY POPULATION
☐ CALM ☐ NERVOUS ☐ AGGRESSIVE	☐ LOW ☐ NORMAL ☐ CROWDED

BROOD PATTERN		EGGS		LARVAE	
TIGHT	SPOTTY	YES	NO	YES	NO

FOOD STORES

HONEY	HIGH	AVERAGE	LOW	NEAR BROOD
POLLEN	HIGH	AVERAGE	LOW	NEAR BROOD

QUEEN OBSERVATION

YES ☐ NO ☐	MARKED- YES ☐ NO ☐	COLOR- W Y R G B

QUEENS CELLS PRESENT		CAPPED		UNCAPPED	
YES	NO	YES	NO	YES	NO

CELL

EMERGENCY	SWARM	SUPERSEDURE

PESTS AND DISEASE

VARROA	ANTS	WAX MOTH
MITES	STARVATION	MOLD
DYSENTARY	SMALL HIVE BEETE	BOLD BROOD
ODOR	BUCKSHOT BROOD	DEAD BEES
OTHER	OTHER	OTHER

TREATMENT	FEEDINGS

TO DO/NOTES	SUPPLIES

DATE	TIME	HIVE ID NUMBER
COLONY NAME		YARD

WEATHER CONDITIONS

☀️ ⛅ ☁️ 🌧️ ⛈️ ❄️
◯ ◯ ◯ ◯ ◯ ◯

HOW MUCH

HONEY	BROOD	SPACE

COLONY TEMPERAMENT	COLONY POPULATION
☐ CALM ☐ NERVOUS ☐ AGGRESSIVE	☐ LOW ☐ NORMAL ☐ CROWDED

BROOD PATTERN		EGGS		LARVAE	
TIGHT	SPOTTY	YES	NO	YES	NO

FOOD STORES

HONEY	HIGH	AVERAGE	LOW	NEAR BROOD
POLLEN	HIGH	AVERAGE	LOW	NEAR BROOD

QUEEN OBSERVATION

YES ☐ NO ☐	MARKED- YES ☐ NO ☐	COLOR- W Y R G B

QUEENS CELLS PRESENT		CAPPED		UNCAPPED	
YES	NO	YES	NO	YES	NO

CELL

EMERGENCY	SWARM	SUPERSEDURE

PESTS AND DISEASE

VARROA	ANTS	WAX MOTH
MITES	STARVATION	MOLD
DYSENTARY	SMALL HIVE BEETE	BOLD BROOD
ODOR	BUCKSHOT BROOD	DEAD BEES
OTHER	OTHER	OTHER

TREATMENT	FEEDINGS

TO DO/NOTES	SUPPLIES

DATE	TIME	HIVE ID NUMBER
COLONY NAME		YARD

WEATHER CONDITIONS

☀️ ☁️ ☁️ 🌧️ ⛈️ ❄️
☐ ☐ ☐ ☐ ☐ ☐

HOW MUCH

HONEY	BROOD	SPACE

COLONY TEMPERAMENT
☐ CALM ☐ NERVOUS ☐ AGGRESSIVE

COLONY POPULATION
☐ LOW ☐ NORMAL ☐ CROWDED

BROOD PATTERN		EGGS		LARVAE	
TIGHT	SPOTTY	YES	NO	YES	NO

FOOD STORES

HONEY	HIGH	AVERAGE	LOW	NEAR BROOD
POLLEN	HIGH	AVERAGE	LOW	NEAR BROOD

QUEEN OBSERVATION

YES ☐   NO ☐   MARKED- YES ☐ NO ☐   COLOR- W Y R G B

QUEENS CELLS PRESENT		CAPPED		UNCAPPED	
YES	NO	YES	NO	YES	NO

CELL

EMERGENCY	SWARM	SUPERSEDURE

PESTS AND DISEASE

VARROA	ANTS	WAX MOTH
MITES	STARVATION	MOLD
DYSENTARY	SMALL HIVE BEETE	BOLD BROOD
ODOR	BUCKSHOT BROOD	DEAD BEES
OTHER	OTHER	OTHER

TREATMENT	FEEDINGS

TO DO/NOTES	SUPPLIES

DATE	TIME	HIVE ID NUMBER
COLONY NAME		YARD

WEATHER CONDITIONS

☀ ○   ⛅ ○   ☁ ○   🌧 ○   ⛈ ○   ❄ ○

HOW MUCH

HONEY	BROOD	SPACE

COLONY TEMPERAMENT	COLONY POPULATION
☐ CALM ☐ NERVOUS ☐ AGGRESSIVE	☐ LOW ☐ NORMAL ☐ CROWDED

BROOD PATTERN		EGGS		LARVAE	
TIGHT	SPOTTY	YES	NO	YES	NO

FOOD STORES

HONEY	HIGH	AVERAGE	LOW	NEAR BROOD
POLLEN	HIGH	AVERAGE	LOW	NEAR BROOD

QUEEN OBSERVATION

YES ☐   NO ☐   MARKED- YES ☐ NO ☐   COLOR- W Y R G B

QUEENS CELLS PRESENT		CAPPED		UNCAPPED	
YES	NO	YES	NO	YES	NO

CELL

EMERGENCY	SWARM	SUPERSEDURE

PESTS AND DISEASE

VARROA	ANTS	WAX MOTH
MITES	STARVATION	MOLD
DYSENTARY	SMALL HIVE BEETE	BOLD BROOD
ODOR	BUCKSHOT BROOD	DEAD BEES
OTHER	OTHER	OTHER

TREATMENT	FEEDINGS

TO DO/NOTES	SUPPLIES

DATE	TIME	HIVE ID NUMBER
COLONY NAME		YARD

WEATHER CONDITIONS

☀️ ☁️ ☁️ 🌧️ ⛈️ ❄️
☐ ☐ ☐ ☐ ☐ ☐

HOW MUCH

HONEY	BROOD	SPACE

COLONY TEMPERAMENT	COLONY POPULATION
☐ CALM ☐ NERVOUS ☐ AGGRESSIVE	☐ LOW ☐ NORMAL ☐ CROWDED

BROOD PATTERN		EGGS		LARVAE	
TIGHT	SPOTTY	YES	NO	YES	NO

FOOD STORES

HONEY	HIGH	AVERAGE	LOW	NEAR BROOD
POLLEN	HIGH	AVERAGE	LOW	NEAR BROOD

QUEEN OBSERVATION

YES ☐ NO ☐	MARKED- YES ☐ NO ☐	COLOR- W Y R G B

QUEENS CELLS PRESENT		CAPPED		UNCAPPED	
YES	NO	YES	NO	YES	NO

CELL

EMERGENCY	SWARM	SUPERSEDURE

PESTS AND DISEASE

VARROA	ANTS	WAX MOTH
MITES	STARVATION	MOLD
DYSENTARY	SMALL HIVE BEETE	BOLD BROOD
ODOR	BUCKSHOT BROOD	DEAD BEES
OTHER	OTHER	OTHER

TREATMENT	FEEDINGS

TO DO/NOTES	SUPPLIES

DATE	TIME	HIVE ID NUMBER
COLONY NAME		YARD

WEATHER CONDITIONS

☀ ☁ ☁☁ 🌧 ⛈ ❄
☐ ☐ ☐ ☐ ☐ ☐

HOW MUCH

HONEY	BROOD	SPACE

COLONY TEMPERAMENT
☐ CALM ☐ NERVOUS ☐ AGGRESSIVE

COLONY POPULATION
☐ LOW ☐ NORMAL ☐ CROWDED

BROOD PATTERN		EGGS		LARVAE	
TIGHT	SPOTTY	YES	NO	YES	NO

FOOD STORES

HONEY	HIGH	AVERAGE	LOW	NEAR BROOD
POLLEN	HIGH	AVERAGE	LOW	NEAR BROOD

QUEEN OBSERVATION

YES ☐ NO ☐   MARKED- YES ☐ NO ☐   COLOR- [W] [Y] [R] [G] [B]

QUEENS CELLS PRESENT		CAPPED		UNCAPPED	
YES	NO	YES	NO	YES	NO

CELL

EMERGENCY	SWARM	SUPERSEDURE

PESTS AND DISEASE

VARROA	ANTS	WAX MOTH
MITES	STARVATION	MOLD
DYSENTARY	SMALL HIVE BEETE	BOLD BROOD
ODOR	BUCKSHOT BROOD	DEAD BEES
OTHER	OTHER	OTHER

TREATMENT	FEEDINGS

TO DO/NOTES	SUPPLIES

DATE	TIME	HIVE ID NUMBER
COLONY NAME		YARD

WEATHER CONDITIONS

☀️ ☁️ ☁️ 🌧️ ⛈️ ❄️
☐ ☐ ☐ ☐ ☐ ☐

HOW MUCH

HONEY	BROOD	SPACE

COLONY TEMPERAMENT	COLONY POPULATION
☐ CALM ☐ NERVOUS ☐ AGGRESSIVE	☐ LOW ☐ NORMAL ☐ CROWDED

BROOD PATTERN		EGGS		LARVAE	
TIGHT	SPOTTY	YES	NO	YES	NO

FOOD STORES

HONEY	HIGH	AVERAGE	LOW	NEAR BROOD
POLLEN	HIGH	AVERAGE	LOW	NEAR BROOD

QUEEN OBSERVATION

YES ☐ NO ☐   MARKED- YES ☐ NO ☐   COLOR- W Y R G B

QUEENS CELLS PRESENT		CAPPED		UNCAPPED	
YES	NO	YES	NO	YES	NO

CELL

EMERGENCY	SWARM	SUPERSEDURE

PESTS AND DISEASE

VARROA	ANTS	WAX MOTH
MITES	STARVATION	MOLD
DYSENTARY	SMALL HIVE BEETE	BOLD BROOD
ODOR	BUCKSHOT BROOD	DEAD BEES
OTHER	OTHER	OTHER

TREATMENT	FEEDINGS

TO DO/NOTES	SUPPLIES

DATE	TIME	HIVE ID NUMBER
COLONY NAME		YARD

WEATHER CONDITIONS

☀️ ⛅ ☁️ 🌧️ ⛈️ ❄️
○ ○ ○ ○ ○ ○

HOW MUCH

HONEY	BROOD	SPACE

COLONY TEMPERAMENT
☐ CALM ☐ NERVOUS ☐ AGGRESSIVE

COLONY POPULATION
☐ LOW ☐ NORMAL ☐ CROWDED

BROOD PATTERN		EGGS		LARVAE	
TIGHT	SPOTTY	YES	NO	YES	NO

FOOD STORES

HONEY	HIGH	AVERAGE	LOW	NEAR BROOD
POLLEN	HIGH	AVERAGE	LOW	NEAR BROOD

QUEEN OBSERVATION

YES ☐ NO ☐ MARKED- YES ☐ NO ☐ COLOR- ☐W ☐Y ☐R ☐G ☐B

QUEENS CELLS PRESENT		CAPPED		UNCAPPED	
YES	NO	YES	NO	YES	NO

CELL

EMERGENCY	SWARM	SUPERSEDURE

PESTS AND DISEASE

VARROA	ANTS	WAX MOTH
MITES	STARVATION	MOLD
DYSENTARY	SMALL HIVE BEETE	BOLD BROOD
ODOR	BUCKSHOT BROOD	DEAD BEES
OTHER	OTHER	OTHER

TREATMENT	FEEDINGS

TO DO/NOTES	SUPPLIES

DATE	TIME	HIVE ID NUMBER
COLONY NAME		YARD

WEATHER CONDITIONS

☀ ☁ ☁ 🌧 ⛈ ❄
☐ ☐ ☐ ☐ ☐ ☐

HOW MUCH

HONEY	BROOD	SPACE

COLONY TEMPERAMENT	COLONY POPULATION
☐ CALM ☐ NERVOUS ☐ AGGRESSIVE	☐ LOW ☐ NORMAL ☐ CROWDED

BROOD PATTERN		EGGS		LARVAE	
TIGHT	SPOTTY	YES	NO	YES	NO

FOOD STORES

HONEY	HIGH	AVERAGE	LOW	NEAR BROOD
POLLEN	HIGH	AVERAGE	LOW	NEAR BROOD

QUEEN OBSERVATION

YES ☐ NO ☐   MARKED- YES ☐ NO ☐   COLOR- W Y R G B

QUEENS CELLS PRESENT		CAPPED		UNCAPPED	
YES	NO	YES	NO	YES	NO

CELL

EMERGENCY	SWARM	SUPERSEDURE

PESTS AND DISEASE

VARROA	ANTS	WAX MOTH
MITES	STARVATION	MOLD
DYSENTARY	SMALL HIVE BEETE	BOLD BROOD
ODOR	BUCKSHOT BROOD	DEAD BEES
OTHER	OTHER	OTHER

TREATMENT	FEEDINGS

TO DO/NOTES	SUPPLIES

DATE	TIME	HIVE ID NUMBER
COLONY NAME		YARD

WEATHER CONDITIONS

☀️ ⛅ ☁️ 🌧️ ⛈️ ❄️
☐ ☐ ☐ ☐ ☐ ☐

HOW MUCH

HONEY	BROOD	SPACE

COLONY TEMPERAMENT	COLONY POPULATION
☐ CALM ☐ NERVOUS ☐ AGGRESSIVE	☐ LOW ☐ NORMAL ☐ CROWDED

BROOD PATTERN		EGGS		LARVAE	
TIGHT	SPOTTY	YES	NO	YES	NO

FOOD STORES

HONEY	HIGH	AVERAGE	LOW	NEAR BROOD
POLLEN	HIGH	AVERAGE	LOW	NEAR BROOD

QUEEN OBSERVATION

YES ☐ NO ☐   MARKED- YES ☐ NO ☐   COLOR- W Y R G B

QUEENS CELLS PRESENT		CAPPED		UNCAPPED	
YES	NO	YES	NO	YES	NO

CELL

EMERGENCY	SWARM	SUPERSEDURE

PESTS AND DISEASE

VARROA	ANTS	WAX MOTH
MITES	STARVATION	MOLD
DYSENTARY	SMALL HIVE BEETE	BOLD BROOD
ODOR	BUCKSHOT BROOD	DEAD BEES
OTHER	OTHER	OTHER

TREATMENT	FEEDINGS

TO DO/NOTES	SUPPLIES

DATE	TIME	HIVE ID NUMBER
COLONY NAME		YARD

WEATHER CONDITIONS

☀️ ☁️ ☁️ 🌧️ ⛈️ ❄️
☐ ☐ ☐ ☐ ☐ ☐

HOW MUCH

HONEY	BROOD	SPACE

COLONY TEMPERAMENT	COLONY POPULATION
☐ CALM ☐ NERVOUS ☐ AGGRESSIVE	☐ LOW ☐ NORMAL ☐ CROWDED

BROOD PATTERN		EGGS		LARVAE	
TIGHT	SPOTTY	YES	NO	YES	NO

FOOD STORES

HONEY	HIGH	AVERAGE	LOW	NEAR BROOD
POLLEN	HIGH	AVERAGE	LOW	NEAR BROOD

QUEEN OBSERVATION

YES ☐ NO ☐   MARKED- YES ☐ NO ☐   COLOR- W ☐ Y ☐ R ☐ G ☐ B ☐

QUEENS CELLS PRESENT		CAPPED		UNCAPPED	
YES	NO	YES	NO	YES	NO

CELL

EMERGENCY	SWARM	SUPERSEDURE

PESTS AND DISEASE

VARROA	ANTS	WAX MOTH
MITES	STARVATION	MOLD
DYSENTARY	SMALL HIVE BEETE	BOLD BROOD
ODOR	BUCKSHOT BROOD	DEAD BEES
OTHER	OTHER	OTHER

TREATMENT	FEEDINGS

TO DO/NOTES	SUPPLIES

DATE		TIME	HIVE ID NUMBER
COLONY NAME		YARD	

WEATHER CONDITIONS

☀ ⛅ ☁ 🌧 ⛈ ❄
○ ○ ○ ○ ○ ○

HOW MUCH

HONEY	BROOD	SPACE

COLONY TEMPERAMENT	COLONY POPULATION
☐ CALM ☐ NERVOUS ☐ AGGRESSIVE	☐ LOW ☐ NORMAL ☐ CROWDED

BROOD PATTERN		EGGS		LARVAE	
TIGHT	SPOTTY	YES	NO	YES	NO

FOOD STORES

HONEY	HIGH	AVERAGE	LOW	NEAR BROOD
POLLEN	HIGH	AVERAGE	LOW	NEAR BROOD

QUEEN OBSERVATION

YES ☐ NO ☐	MARKED- YES ☐ NO ☐	COLOR- W Y R G B

QUEENS CELLS PRESENT		CAPPED		UNCAPPED	
YES	NO	YES	NO	YES	NO

CELL

EMERGENCY	SWARM	SUPERSEDURE

PESTS AND DISEASE

VARROA	ANTS	WAX MOTH
MITES	STARVATION	MOLD
DYSENTARY	SMALL HIVE BEETE	BOLD BROOD
ODOR	BUCKSHOT BROOD	DEAD BEES
OTHER	OTHER	OTHER

TREATMENT	FEEDINGS

TO DO/NOTES	SUPPLIES

DATE	TIME	HIVE ID NUMBER
COLONY NAME		YARD

WEATHER CONDITIONS

☀️ ☐ ⛅ ☐ ☁️ ☐ 🌧️ ☐ ⛈️ ☐ ❄️ ☐

HOW MUCH

HONEY	BROOD	SPACE

COLONY TEMPERAMENT	COLONY POPULATION
☐ CALM ☐ NERVOUS ☐ AGGRESSIVE	☐ LOW ☐ NORMAL ☐ CROWDED

BROOD PATTERN		EGGS		LARVAE	
TIGHT	SPOTTY	YES	NO	YES	NO

FOOD STORES

HONEY	HIGH	AVERAGE	LOW	NEAR BROOD
POLLEN	HIGH	AVERAGE	LOW	NEAR BROOD

QUEEN OBSERVATION

YES ☐   NO ☐   MARKED- YES ☐ NO ☐   COLOR- ☒W ☐Y ☐R ☐G ☐B

QUEENS CELLS PRESENT		CAPPED		UNCAPPED	
YES	NO	YES	NO	YES	NO

CELL

EMERGENCY	SWARM	SUPERSEDURE

PESTS AND DISEASE

VARROA	ANTS	WAX MOTH
MITES	STARVATION	MOLD
DYSENTARY	SMALL HIVE BEETE	BOLD BROOD
ODOR	BUCKSHOT BROOD	DEAD BEES
OTHER	OTHER	OTHER

TREATMENT	FEEDINGS

TO DO/NOTES	SUPPLIES

DATE		TIME	HIVE ID NUMBER
COLONY NAME		YARD	

WEATHER CONDITIONS

☀️ ○   ⛅ ○   ☁️ ○   🌧️ ○   ⛈️ ○   ❄️ ○

HOW MUCH

HONEY	BROOD	SPACE

COLONY TEMPERAMENT	COLONY POPULATION
☐ CALM ☐ NERVOUS ☐ AGGRESSIVE	☐ LOW ☐ NORMAL ☐ CROWDED

BROOD PATTERN		EGGS		LARVAE	
TIGHT	SPOTTY	YES	NO	YES	NO

FOOD STORES

HONEY	HIGH	AVERAGE	LOW	NEAR BROOD
POLLEN	HIGH	AVERAGE	LOW	NEAR BROOD

QUEEN OBSERVATION

YES ☐   NO ☐   MARKED- YES ☐   NO ☐   COLOR- W Y R G B

QUEENS CELLS PRESENT		CAPPED		UNCAPPED	
YES	NO	YES	NO	YES	NO

CELL

EMERGENCY	SWARM	SUPERSEDURE

PESTS AND DISEASE

VARROA	ANTS	WAX MOTH
MITES	STARVATION	MOLD
DYSENTARY	SMALL HIVE BEETE	BOLD BROOD
ODOR	BUCKSHOT BROOD	DEAD BEES
OTHER	OTHER	OTHER

TREATMENT	FEEDINGS

TO DO/NOTES	SUPPLIES

DATE	TIME	HIVE ID NUMBER
COLONY NAME		YARD

WEATHER CONDITIONS

☀️ ⛅ ☁️ 🌧️ ⛈️ ❄️
☐ ☐ ☐ ☐ ☐ ☐

HOW MUCH

HONEY	BROOD	SPACE

COLONY TEMPERAMENT	COLONY POPULATION
☐ CALM ☐ NERVOUS ☐ AGGRESSIVE	☐ LOW ☐ NORMAL ☐ CROWDED

BROOD PATTERN		EGGS		LARVAE	
TIGHT	SPOTTY	YES	NO	YES	NO

FOOD STORES

HONEY	HIGH	AVERAGE	LOW	NEAR BROOD
POLLEN	HIGH	AVERAGE	LOW	NEAR BROOD

QUEEN OBSERVATION

YES ☐   NO ☐   MARKED- YES ☐ NO ☐   COLOR- W Y R G B

QUEENS CELLS PRESENT		CAPPED		UNCAPPED	
YES	NO	YES	NO	YES	NO

CELL

EMERGENCY	SWARM	SUPERSEDURE

PESTS AND DISEASE

VARROA	ANTS	WAX MOTH
MITES	STARVATION	MOLD
DYSENTARY	SMALL HIVE BEETE	BOLD BROOD
ODOR	BUCKSHOT BROOD	DEAD BEES
OTHER	OTHER	OTHER

TREATMENT	FEEDINGS

TO DO/NOTES	SUPPLIES

DATE	TIME	HIVE ID NUMBER
COLONY NAME		YARD

WEATHER CONDITIONS

☀️ ○   ⛅ ○   ☁️ ○   🌧️ ○   ⛈️ ○   ❄️ ○

HOW MUCH

HONEY	BROOD	SPACE

COLONY TEMPERAMENT	COLONY POPULATION
☐ CALM ☐ NERVOUS ☐ AGGRESSIVE	☐ LOW ☐ NORMAL ☐ CROWDED

BROOD PATTERN		EGGS		LARVAE	
TIGHT	SPOTTY	YES	NO	YES	NO

FOOD STORES

HONEY	HIGH	AVERAGE	LOW	NEAR BROOD
POLLEN	HIGH	AVERAGE	LOW	NEAR BROOD

QUEEN OBSERVATION

YES ☐   NO ☐   MARKED- YES ☐   NO ☐   COLOR- W Y R G B

QUEENS CELLS PRESENT		CAPPED		UNCAPPED	
YES	NO	YES	NO	YES	NO

CELL

EMERGENCY	SWARM	SUPERSEDURE

PESTS AND DISEASE

VARROA	ANTS	WAX MOTH
MITES	STARVATION	MOLD
DYSENTARY	SMALL HIVE BEETE	BOLD BROOD
ODOR	BUCKSHOT BROOD	DEAD BEES
OTHER	OTHER	OTHER

TREATMENT	FEEDINGS

TO DO/NOTES	SUPPLIES

DATE	TIME	HIVE ID NUMBER
COLONY NAME		YARD

WEATHER CONDITIONS

☀️ ☁️ ☁️ 🌧️ ⛈️ ❄️
☐ ☐ ☐ ☐ ☐ ☐

HOW MUCH

HONEY	BROOD	SPACE

COLONY TEMPERAMENT	COLONY POPULATION
☐ CALM ☐ NERVOUS ☐ AGGRESSIVE	☐ LOW ☐ NORMAL ☐ CROWDED

BROOD PATTERN		EGGS		LARVAE	
TIGHT	SPOTTY	YES	NO	YES	NO

FOOD STORES

HONEY	HIGH	AVERAGE	LOW	NEAR BROOD
POLLEN	HIGH	AVERAGE	LOW	NEAR BROOD

QUEEN OBSERVATION

YES ☐ NO ☐   MARKED- YES ☐ NO ☐   COLOR- W Y R G B

QUEENS CELLS PRESENT		CAPPED		UNCAPPED	
YES	NO	YES	NO	YES	NO

CELL

EMERGENCY	SWARM	SUPERSEDURE

PESTS AND DISEASE

VARROA	ANTS	WAX MOTH
MITES	STARVATION	MOLD
DYSENTARY	SMALL HIVE BEETE	BOLD BROOD
ODOR	BUCKSHOT BROOD	DEAD BEES
OTHER	OTHER	OTHER

TREATMENT	FEEDINGS

TO DO/NOTES	SUPPLIES

DATE		TIME	HIVE ID NUMBER
COLONY NAME		YARD	

WEATHER CONDITIONS

☀ ⛅ ☁ 🌧 ⛈ ❄
○ ○ ○ ○ ○ ○

HOW MUCH

HONEY	BROOD	SPACE

COLONY TEMPERAMENT	COLONY POPULATION
☐ CALM ☐ NERVOUS ☐ AGGRESSIVE	☐ LOW ☐ NORMAL ☐ CROWDED

BROOD PATTERN		EGGS		LARVAE	
TIGHT	SPOTTY	YES	NO	YES	NO

FOOD STORES

HONEY	HIGH	AVERAGE	LOW	NEAR BROOD
POLLEN	HIGH	AVERAGE	LOW	NEAR BROOD

QUEEN OBSERVATION

YES ☐ NO ☐   MARKED- YES ☐ NO ☐   COLOR- W Y R G B

QUEENS CELLS PRESENT		CAPPED		UNCAPPED	
YES	NO	YES	NO	YES	NO

CELL

EMERGENCY	SWARM	SUPERSEDURE

PESTS AND DISEASE

VARROA	ANTS	WAX MOTH
MITES	STARVATION	MOLD
DYSENTARY	SMALL HIVE BEETE	BOLD BROOD
ODOR	BUCKSHOT BROOD	DEAD BEES
OTHER	OTHER	OTHER

TREATMENT	FEEDINGS

TO DO/NOTES	SUPPLIES

DATE	TIME	HIVE ID NUMBER
COLONY NAME		YARD

WEATHER CONDITIONS

☀️ ⛅ ☁️ 🌧️ ⛈️ ❄️
☐ ☐ ☐ ☐ ☐ ☐

HOW MUCH

HONEY	BROOD	SPACE

COLONY TEMPERAMENT	COLONY POPULATION
☐ CALM ☐ NERVOUS ☐ AGGRESSIVE	☐ LOW ☐ NORMAL ☐ CROWDED

BROOD PATTERN		EGGS		LARVAE	
TIGHT	SPOTTY	YES	NO	YES	NO

FOOD STORES

HONEY	HIGH	AVERAGE	LOW	NEAR BROOD
POLLEN	HIGH	AVERAGE	LOW	NEAR BROOD

QUEEN OBSERVATION

YES ☐   NO ☐   MARKED- YES ☐   NO ☐   COLOR- W Y R G B

QUEENS CELLS PRESENT		CAPPED		UNCAPPED	
YES	NO	YES	NO	YES	NO

CELL

EMERGENCY	SWARM	SUPERSEDURE

PESTS AND DISEASE

VARROA	ANTS	WAX MOTH
MITES	STARVATION	MOLD
DYSENTARY	SMALL HIVE BEETE	BOLD BROOD
ODOR	BUCKSHOT BROOD	DEAD BEES
OTHER	OTHER	OTHER

TREATMENT	FEEDINGS

TO DO/NOTES	SUPPLIES

DATE	TIME	HIVE ID NUMBER
COLONY NAME		YARD

WEATHER CONDITIONS

☀️ ⛅ ☁️ 🌧️ ⛈️ ❄️
○ ○ ○ ○ ○ ○

HOW MUCH

HONEY	BROOD	SPACE

COLONY TEMPERAMENT	COLONY POPULATION
☐ CALM ☐ NERVOUS ☐ AGGRESSIVE	☐ LOW ☐ NORMAL ☐ CROWDED

BROOD PATTERN		EGGS		LARVAE	
TIGHT	SPOTTY	YES	NO	YES	NO

FOOD STORES

HONEY	HIGH	AVERAGE	LOW	NEAR BROOD
POLLEN	HIGH	AVERAGE	LOW	NEAR BROOD

QUEEN OBSERVATION

YES ☐ NO ☐   MARKED- YES ☐ NO ☐   COLOR- W Y R G B

QUEENS CELLS PRESENT		CAPPED		UNCAPPED	
YES	NO	YES	NO	YES	NO

CELL

EMERGENCY	SWARM	SUPERSEDURE

PESTS AND DISEASE

VARROA	ANTS	WAX MOTH
MITES	STARVATION	MOLD
DYSENTARY	SMALL HIVE BEETE	BOLD BROOD
ODOR	BUCKSHOT BROOD	DEAD BEES
OTHER	OTHER	OTHER

TREATMENT	FEEDINGS

TO DO/NOTES	SUPPLIES

DATE	TIME	HIVE ID NUMBER
COLONY NAME	YARD	

WEATHER CONDITIONS

☀️ ⛅ ☁️ 🌧️ ⛈️ ❄️
☐ ☐ ☐ ☐ ☐ ☐

HOW MUCH

HONEY	BROOD	SPACE

COLONY TEMPERAMENT	COLONY POPULATION
☐ CALM ☐ NERVOUS ☐ AGGRESSIVE	☐ LOW ☐ NORMAL ☐ CROWDED

BROOD PATTERN		EGGS		LARVAE	
TIGHT	SPOTTY	YES	NO	YES	NO

FOOD STORES

HONEY	HIGH	AVERAGE	LOW	NEAR BROOD
POLLEN	HIGH	AVERAGE	LOW	NEAR BROOD

QUEEN OBSERVATION

YES ☐ NO ☐   MARKED- YES ☐ NO ☐   COLOR- [W] [Y] [R] [G] [B]

QUEENS CELLS PRESENT		CAPPED		UNCAPPED	
YES	NO	YES	NO	YES	NO

CELL

EMERGENCY	SWARM	SUPERSEDURE

PESTS AND DISEASE

VARROA	ANTS	WAX MOTH
MITES	STARVATION	MOLD
DYSENTARY	SMALL HIVE BEETE	BOLD BROOD
ODOR	BUCKSHOT BROOD	DEAD BEES
OTHER	OTHER	OTHER

TREATMENT	FEEDINGS

TO DO/NOTES	SUPPLIES

DATE		TIME	HIVE ID NUMBER
COLONY NAME		YARD	

WEATHER CONDITIONS

☀ ⛅ ☁ 🌧 ⛈ ❄
○ ○ ○ ○ ○ ○

HOW MUCH

HONEY	BROOD	SPACE

COLONY TEMPERAMENT	COLONY POPULATION
☐ CALM ☐ NERVOUS ☐ AGGRESSIVE	☐ LOW ☐ NORMAL ☐ CROWDED

BROOD PATTERN		EGGS		LARVAE	
TIGHT	SPOTTY	YES	NO	YES	NO

FOOD STORES

HONEY	HIGH	AVERAGE	LOW	NEAR BROOD
POLLEN	HIGH	AVERAGE	LOW	NEAR BROOD

QUEEN OBSERVATION

YES ☐ NO ☐   MARKED- YES ☐ NO ☐   COLOR- W Y R G B

QUEENS CELLS PRESENT		CAPPED		UNCAPPED	
YES	NO	YES	NO	YES	NO

CELL

EMERGENCY	SWARM	SUPERSEDURE

PESTS AND DISEASE

VARROA	ANTS	WAX MOTH
MITES	STARVATION	MOLD
DYSENTARY	SMALL HIVE BEETE	BOLD BROOD
ODOR	BUCKSHOT BROOD	DEAD BEES
OTHER	OTHER	OTHER

TREATMENT	FEEDINGS

TO DO/NOTES	SUPPLIES

DATE	TIME	HIVE ID NUMBER
COLONY NAME		YARD

WEATHER CONDITIONS

☀️ ⛅ ☁️ 🌧️ ⛈️ ❄️
☐ ☐ ☐ ☐ ☐ ☐

HOW MUCH

HONEY	BROOD	SPACE

COLONY TEMPERAMENT	COLONY POPULATION
☐ CALM ☐ NERVOUS ☐ AGGRESSIVE	☐ LOW ☐ NORMAL ☐ CROWDED

BROOD PATTERN		EGGS		LARVAE	
TIGHT	SPOTTY	YES	NO	YES	NO

FOOD STORES

HONEY	HIGH	AVERAGE	LOW	NEAR BROOD
POLLEN	HIGH	AVERAGE	LOW	NEAR BROOD

QUEEN OBSERVATION

YES ☐ NO ☐	MARKED- YES ☐ NO ☐	COLOR- W Y R G B

QUEENS CELLS PRESENT		CAPPED		UNCAPPED	
YES	NO	YES	NO	YES	NO

CELL

EMERGENCY	SWARM	SUPERSEDURE

PESTS AND DISEASE

VARROA	ANTS	WAX MOTH
MITES	STARVATION	MOLD
DYSENTARY	SMALL HIVE BEETE	BOLD BROOD
ODOR	BUCKSHOT BROOD	DEAD BEES
OTHER	OTHER	OTHER

TREATMENT	FEEDINGS

TO DO/NOTES	SUPPLIES

DATE	TIME	HIVE ID NUMBER
COLONY NAME	YARD	

WEATHER CONDITIONS

☀️ ○   ⛅ ○   ☁️ ○   🌧️ ○   ⛈️ ○   ❄️ ○

HOW MUCH

HONEY	BROOD	SPACE

COLONY TEMPERAMENT	COLONY POPULATION
☐ CALM ☐ NERVOUS ☐ AGGRESSIVE	☐ LOW ☐ NORMAL ☐ CROWDED

BROOD PATTERN		EGGS		LARVAE	
TIGHT	SPOTTY	YES	NO	YES	NO

FOOD STORES

HONEY	HIGH	AVERAGE	LOW	NEAR BROOD
POLLEN	HIGH	AVERAGE	LOW	NEAR BROOD

QUEEN OBSERVATION

YES ☐   NO ☐   MARKED- YES ☐   NO ☐   COLOR- W Y R G B

QUEENS CELLS PRESENT		CAPPED		UNCAPPED	
YES	NO	YES	NO	YES	NO

CELL

EMERGENCY	SWARM	SUPERSEDURE

PESTS AND DISEASE

VARROA	ANTS	WAX MOTH
MITES	STARVATION	MOLD
DYSENTARY	SMALL HIVE BEETE	BOLD BROOD
ODOR	BUCKSHOT BROOD	DEAD BEES
OTHER	OTHER	OTHER

TREATMENT	FEEDINGS

TO DO/NOTES	SUPPLIES

DATE	TIME	HIVE ID NUMBER
COLONY NAME		YARD

WEATHER CONDITIONS

☀️	⛅	☁️	🌧️	⛈️	❄️
☐	☐	☐	☐	☐	☐

HOW MUCH

HONEY	BROOD	SPACE

COLONY TEMPERAMENT	COLONY POPULATION
☐ CALM ☐ NERVOUS ☐ AGGRESSIVE	☐ LOW ☐ NORMAL ☐ CROWDED

BROOD PATTERN		EGGS		LARVAE	
TIGHT	SPOTTY	YES	NO	YES	NO

FOOD STORES

HONEY	HIGH	AVERAGE	LOW	NEAR BROOD
POLLEN	HIGH	AVERAGE	LOW	NEAR BROOD

QUEEN OBSERVATION

YES ☐ NO ☐	MARKED- YES ☐ NO ☐	COLOR- W Y R G B

QUEENS CELLS PRESENT		CAPPED		UNCAPPED	
YES	NO	YES	NO	YES	NO

CELL

EMERGENCY	SWARM	SUPERSEDURE

PESTS AND DISEASE

VARROA	ANTS	WAX MOTH
MITES	STARVATION	MOLD
DYSENTARY	SMALL HIVE BEETE	BOLD BROOD
ODOR	BUCKSHOT BROOD	DEAD BEES
OTHER	OTHER	OTHER

TREATMENT	FEEDINGS

TO DO/NOTES	SUPPLIES

DATE	TIME	HIVE ID NUMBER
COLONY NAME		YARD

WEATHER CONDITIONS

☀ ☁ ☁☁ 🌧 ⛈ ❄
◯ ◯ ◯ ◯ ◯ ◯

HOW MUCH

HONEY	BROOD	SPACE

COLONY TEMPERAMENT	COLONY POPULATION
☐ CALM ☐ NERVOUS ☐ AGGRESSIVE	☐ LOW ☐ NORMAL ☐ CROWDED

BROOD PATTERN		EGGS		LARVAE	
TIGHT	SPOTTY	YES	NO	YES	NO

FOOD STORES

HONEY	HIGH	AVERAGE	LOW	NEAR BROOD
POLLEN	HIGH	AVERAGE	LOW	NEAR BROOD

QUEEN OBSERVATION

YES ☐ NO ☐   MARKED- YES ☐ NO ☐   COLOR- [W] [Y] [R] [G] [B]

QUEENS CELLS PRESENT		CAPPED		UNCAPPED	
YES	NO	YES	NO	YES	NO

CELL

EMERGENCY	SWARM	SUPERSEDURE

PESTS AND DISEASE

VARROA	ANTS	WAX MOTH
MITES	STARVATION	MOLD
DYSENTARY	SMALL HIVE BEETE	BOLD BROOD
ODOR	BUCKSHOT BROOD	DEAD BEES
OTHER	OTHER	OTHER

TREATMENT	FEEDINGS

TO DO/NOTES	SUPPLIES

DATE	TIME	HIVE ID NUMBER
COLONY NAME	YARD	

WEATHER CONDITIONS

☀️ ☁️ ☁️ 🌧️ ⛈️ ❄️
◯ ◯ ◯ ◯ ◯ ◯

HOW MUCH

HONEY	BROOD	SPACE

COLONY TEMPERAMENT	COLONY POPULATION
☐ CALM ☐ NERVOUS ☐ AGGRESSIVE	☐ LOW ☐ NORMAL ☐ CROWDED

BROOD PATTERN		EGGS		LARVAE	
TIGHT	SPOTTY	YES	NO	YES	NO

FOOD STORES

HONEY	HIGH	AVERAGE	LOW	NEAR BROOD
POLLEN	HIGH	AVERAGE	LOW	NEAR BROOD

QUEEN OBSERVATION

YES ☐   NO ☐   MARKED- YES ☐ NO ☐   COLOR- W Y R G B

QUEENS CELLS PRESENT		CAPPED		UNCAPPED	
YES	NO	YES	NO	YES	NO

CELL

EMERGENCY	SWARM	SUPERSEDURE

PESTS AND DISEASE

VARROA	ANTS	WAX MOTH
MITES	STARVATION	MOLD
DYSENTARY	SMALL HIVE BEETE	BOLD BROOD
ODOR	BUCKSHOT BROOD	DEAD BEES
OTHER	OTHER	OTHER

TREATMENT	FEEDINGS

TO DO/NOTES	SUPPLIES

DATE	TIME	HIVE ID NUMBER
COLONY NAME		YARD

WEATHER CONDITIONS

☀️ ⛅ ☁️ 🌧️ ⛈️ ❄️
◯ ◯ ◯ ◯ ◯ ◯

HOW MUCH

HONEY	BROOD	SPACE

COLONY TEMPERAMENT	COLONY POPULATION
☐ CALM ☐ NERVOUS ☐ AGGRESSIVE	☐ LOW ☐ NORMAL ☐ CROWDED

BROOD PATTERN		EGGS		LARVAE	
TIGHT	SPOTTY	YES	NO	YES	NO

FOOD STORES

HONEY	HIGH	AVERAGE	LOW	NEAR BROOD
POLLEN	HIGH	AVERAGE	LOW	NEAR BROOD

QUEEN OBSERVATION

YES ☐ NO ☐	MARKED- YES ☐ NO ☐	COLOR- W Y R G B

QUEENS CELLS PRESENT		CAPPED		UNCAPPED	
YES	NO	YES	NO	YES	NO

CELL

EMERGENCY	SWARM	SUPERSEDURE

PESTS AND DISEASE

VARROA	ANTS	WAX MOTH
MITES	STARVATION	MOLD
DYSENTARY	SMALL HIVE BEETE	BOLD BROOD
ODOR	BUCKSHOT BROOD	DEAD BEES
OTHER	OTHER	OTHER

TREATMENT	FEEDINGS

TO DO/NOTES	SUPPLIES

DATE	TIME	HIVE ID NUMBER
COLONY NAME		YARD

WEATHER CONDITIONS

☀️ ⛅ ☁️ 🌧️ ⛈️ ❄️
☐ ☐ ☐ ☐ ☐ ☐

HOW MUCH

HONEY	BROOD	SPACE

COLONY TEMPERAMENT	COLONY POPULATION
☐ CALM ☐ NERVOUS ☐ AGGRESSIVE	☐ LOW ☐ NORMAL ☐ CROWDED

BROOD PATTERN		EGGS		LARVAE	
TIGHT	SPOTTY	YES	NO	YES	NO

FOOD STORES

HONEY	HIGH	AVERAGE	LOW	NEAR BROOD
POLLEN	HIGH	AVERAGE	LOW	NEAR BROOD

QUEEN OBSERVATION

YES ☐ NO ☐	MARKED- YES ☐ NO ☐	COLOR- W Y R G B

QUEENS CELLS PRESENT		CAPPED		UNCAPPED	
YES	NO	YES	NO	YES	NO

CELL

EMERGENCY	SWARM	SUPERSEDURE

PESTS AND DISEASE

VARROA	ANTS	WAX MOTH
MITES	STARVATION	MOLD
DYSENTARY	SMALL HIVE BEETE	BOLD BROOD
ODOR	BUCKSHOT BROOD	DEAD BEES
OTHER	OTHER	OTHER

TREATMENT	FEEDINGS

TO DO/NOTES	SUPPLIES

DATE	TIME	HIVE ID NUMBER
COLONY NAME		YARD

WEATHER CONDITIONS

☀️ ☁️ ☁️ 🌧️ ⛈️ ❄️
○ ○ ○ ○ ○ ○

HOW MUCH

HONEY	BROOD	SPACE

COLONY TEMPERAMENT	COLONY POPULATION
☐ CALM ☐ NERVOUS ☐ AGGRESSIVE	☐ LOW ☐ NORMAL ☐ CROWDED

BROOD PATTERN		EGGS		LARVAE	
TIGHT	SPOTTY	YES	NO	YES	NO

FOOD STORES

HONEY	HIGH	AVERAGE	LOW	NEAR BROOD
POLLEN	HIGH	AVERAGE	LOW	NEAR BROOD

QUEEN OBSERVATION

YES ☐ NO ☐ MARKED- YES ☐ NO ☐ COLOR- W Y R G B

QUEENS CELLS PRESENT		CAPPED		UNCAPPED	
YES	NO	YES	NO	YES	NO

CELL

EMERGENCY	SWARM	SUPERSEDURE

PESTS AND DISEASE

VARROA	ANTS	WAX MOTH
MITES	STARVATION	MOLD
DYSENTARY	SMALL HIVE BEETE	BOLD BROOD
ODOR	BUCKSHOT BROOD	DEAD BEES
OTHER	OTHER	OTHER

TREATMENT	FEEDINGS

TO DO/NOTES	SUPPLIES

DATE	TIME	HIVE ID NUMBER
COLONY NAME		YARD

WEATHER CONDITIONS

☀️ ⛅ ☁️ 🌧️ ⛈️ ❄️
○ ○ ○ ○ ○ ○

HOW MUCH

HONEY	BROOD	SPACE

COLONY TEMPERAMENT	COLONY POPULATION
☐ CALM ☐ NERVOUS ☐ AGGRESSIVE	☐ LOW ☐ NORMAL ☐ CROWDED

BROOD PATTERN		EGGS		LARVAE	
TIGHT	SPOTTY	YES	NO	YES	NO

FOOD STORES

HONEY	HIGH	AVERAGE	LOW	NEAR BROOD
POLLEN	HIGH	AVERAGE	LOW	NEAR BROOD

QUEEN OBSERVATION

YES ☐ NO ☐	MARKED- YES ☐ NO ☐	COLOR- W Y R G B

QUEENS CELLS PRESENT		CAPPED		UNCAPPED	
YES	NO	YES	NO	YES	NO

CELL

EMERGENCY	SWARM	SUPERSEDURE

PESTS AND DISEASE

VARROA	ANTS	WAX MOTH
MITES	STARVATION	MOLD
DYSENTARY	SMALL HIVE BEETE	BOLD BROOD
ODOR	BUCKSHOT BROOD	DEAD BEES
OTHER	OTHER	OTHER

TREATMENT	FEEDINGS

TO DO/NOTES	SUPPLIES

APIARY SETUP

APIARY SETUP

APIARY SETUP

APIARY SETUP

APIARY SETUP

www.ingramcontent.com/pod-product-compliance
Lightning Source LLC
Chambersburg PA
CBHW081154070526
44583CB00021B/2836